Marion Junior High Library
2 Patriot Drive
Marion, AR 72364

Staying Safe at Work

A familiar work setting may give you a false sense of safety.

THE GET PREPARED LIBRARY

VIOLENCE PREVENTION FOR YOUNG WOMEN

Staying Safe at Work

Donna Chaiet

THE ROSEN PUBLISHING GROUP, INC.
NEW YORK

Introduction

You are alone at the cash register in a retail store, and an armed man approaches and demands the money and your jewelry. What should you do?

A coworker or employer makes comments about your body and starts to touch you in a sexual way. How should you handle the situation?

Whether you are working alone in a store or waiting tables at a crowded restaurant, you are vulnerable to attack. Most of us can relate to fear of being alone on a dark street, but we may not perceive a familiar workplace as having any risk of personal assault. Unfortunately, that is no longer

Every job, even that of dog walker, involves safety risks.

true. Recent data by the U.S. Bureau of Labor Statistics indicate that 40 percent of women who die in the workplace are victims of murder. Between 150 and 180 women are murdered at work per year.

As you get older, being in the workplace will occupy larger portions of your time. The more people with whom you interact on a given day, the greater the likelihood of possible problems. This does not mean that you should not work. It only means that you must take responsibility for your own safety by acknowledging risky behaviors and avoiding or deescalating higher-risk situations. Many of you naturally follow such precautions as avoiding using an Automated Teller Machine (ATM) late at night when you're alone. However, deescalating a dangerous situation takes more skill and does not always come naturally.

Learn how to be aware, assess the situation, and make appropriate choices. With a little practice, you can master some easy skills that will serve you through your teen years and into adulthood.◆

Make a mental checklist of your physical surroundings when you begin a new job.

chapter 1

Awareness

Being aware is the most important safety skill to have. Many of you may think that you have fine-tuned awareness skills. If you do, that is terrific. Check them out by paying attention to small details in your environment. See if you can keep track of how many people are behind you when you are walking to and from school. If you live in a rural setting, see if you can notice alterations such as fallen trees or seasonal changes.

These types of awareness skills can be very important in your work setting as well. When you first begin to work at a new job, you will very likely be in a setting different from your home or school. You may work in a community other than where you live, where the customers, coworkers, and bosses are almost all strangers. Most likely you will also be working with supervisors who have some authority over you and coworkers who are your peers.

On your first day at work, begin to make a mental checklist of the physical surroundings. For example, if you work in a fast-food restaurant, note the entrances and exits. Note if there are security guards and where they stand. Is your workspace small and confined, such as a cashier's position? Or is it spacious, like a sales clerk's in a department store?

Start paying attention to the environment right outside your workplace when you enter or leave. Are there lots of people, or is it quiet? Is it light or dark? What do the entrance and exit look like? Are you outside near parking lots, alleys, sidewalks, trees, and bushes; or indoors as in a mall?

Next, observe the people you work with. Who are they? Adults, teens, senior citizens? Do they share your cultural heritage?

Notice the environment around your workplace.

Do they speak and/or understand English? Pay attention to your supervisor. Does he or she interact with the staff, ignore them, or boss them around? How do the coworkers interact? Do they joke around a lot, smoke, drink, goof off? Are they diligent and quiet, speaking little to each other? The goal is to get a feel for the work environment.

The reason this inventory is important is that you may be able to determine very early on that the work culture compromises your safety. For example, if workers drink or do drugs and make derogatory remarks about women, homosexuals, or people of other ethnicities, you may decide that this is not the kind of place where you want to work. Similarly, if you have to leave work by a back exit late at night and walk through a deserted alley, you may want to reconsider working a night shift or request that you be allowed to leave by a different exit that has better lighting.

The time to take this inventory is when you start working. Once you have been there a while, things may start feeling okay (because they have become familiar), even though they still may compromise your safety. For example, your coworkers' bigoted attitudes may begin to seem tolerable, or the alley may not seem so dark or so empty, even though nothing has changed. People adjust and adapt very quickly, which is a good thing in many circumstances. But you may lose some perspective on the situation. Take the inventory early in your work career and act accordingly if anything seems awry, unsafe, or makes you feel uncomfortable.◆

chapter 2

Get Prepared and Have a Plan

Having inventoried your work environment and determined that it is okay doesn't mean that you are guaranteed perfect safety. Crime occurs in many settings, not just on the streets. The next step is to figure out a safety plan for your particular situation. Many fast-food restaurants and convenience stores are targets for robberies. Most of these places have some sort of security system, including surveillance cameras or a panic button by the register that contacts the police. If they don't have a procedure, ask for one. If they don't provide one for you, work out in your mind what you might do if you were confronted

If you work as a cashier, it is a good idea to figure out a plan of safety in case of robbery.

with a criminal asking for the store's cash register money. Let's look at an example.

Monica

Monica works at a one-hour photo store in the parking lot of a suburban minimall. There is no panic button in the store, even though she works alone and does an exclusively cash business. She has examined her work environment and knows that there is only one door and one window. It is at least 200 yards to the other stores in the minimall. She is pretty sure that if she yelled, no one at the stores could hear her.

One evening at about 5 p.m. a man approached her window, pointed a gun at her, and demanded all the money in the cash register and her jewelry. Monica quickly assessed the situation and realized that she was in serious danger. She also realized, however, that she still had choices and options. The first thing she did was to force herself to breathe and to listen to what he was asking for. She also made a mental checklist of his behavior. He appeared to be nervous. His voice trembled, and his hand was shaking.

Once she got control of her breathing, she calmly said that she was going to cooperate and began to describe to him exactly what she was doing. "Okay. I'm going to go and unlock the cash register. I am going to reach into my pocket and get the key." The last thing Monica wanted was to have this guy think she was reaching for a gun. Monica had also predetermined that her possessions and the money in the register were not worth her life. She wasn't faking; she was really cooperating. Even so, she was not cowering, whining, or begging for her life. Her body language, facial expression, and tone of voice communicated calm confidence. She wanted to keep him as calm as possible and stay calm herself. Sometimes criminals begin a property crime such as a robbery, and then the crime escalates to something more violent, such as beating, raping, or killing the victim. Sudden movements that may look like reaching for a weapon could give the assailant a reason to increase the level of personal violence. Monica did not want to get hurt in this crime.

Monica got all the cash in the register and handed it over. She then very slowly and deliberately took off her watch and earrings and handed them to the robber. Monica also did one last thing. She took a close look at him, paying attention to details like facial hair, scarring and moles, the color of his hair and eyes, and his shoes. It is easy to change clothing or to take off the top layer, but shoes are less likely to be changed. When he left, Monica called the police and told them exactly what had happened.

You may put yourself in more danger if you fight for your possessions rather than giving them up in a robbery. They are replaceable; your life is not.

The safest thing to do during a crime against your personal property or the property of a business is to cooperate and give up the money or goods. This is not the time to be a hero and to fight off the bad guys. Money and property can be replaced; your life cannot. However, you don't want to give the criminal reason to escalate the property crime into a personal crime against you. Monica was very clear about behaving in a cooperative way, not in a way that gave the criminal an opportunity to escalate the crime.◆

chapter 3

Know Your Boundaries

What we know about crime is that it is not limited to crime against property. Particularly for women, crime occurs in more personally violent ways, ranging from assault to rape. Personal crimes don't happen only with strangers; they can (and do) happen with people we know. The United States Department of Justice report, *Violence against Women: A National Crime Victimization Survey Report*, found that over two thirds of violent acts against women were committed by someone known to them.

One of the first steps in learning how to prevent personal crimes from occurring with people we know is to understand the concept called "boundaries." A boundary is a physical line that marks or fixes a limit. Your schoolyard is probably marked off by a fence. That is a boundary. The basketball court in your gym is marked off by painted lines. That is also a boundary.

People have boundaries too. Most of you can probably relate to a physical bubble surrounding your body that friends do not cross unless they give you a hug or a kiss. Our culture and the way we were brought up determine the size of this space bubble or "comfort zone." For most of us, a conversational distance would be about one arm's length, close enough to touch the other person's shoulder, but not close enough to kiss without leaning in.

We also learn physical boundaries from our families. Some families hug and kiss regularly; other families rarely touch each other. Some of us grow up in homes where the boundaries are not very clear. Parents and siblings may think it is okay to enter your bedroom or the bathroom without knocking first. Relatives may assume it is

People have boundaries dictating how close we want other people to be to us. Sometimes these boundaries are violated.

okay to hug you or pinch your cheek without your permission. Often privacy is not respected in other ways, such as a family member's reading a letter addressed to you or reading your diary.

Learning to understand your boundary system is important to personal safety. We may tolerate some behavior because we grew up with it. For example, if you grew up in a home where people hugged you and pinched your cheek, you may believe that it is okay for people to touch you without asking your permission. It is my belief that no one (not even a parent) has the right to touch me without my permission.

It is important to note that there is nothing "right" or "wrong" about your boundary system. It is what it is. What is important is that your beliefs about your "comfort zone" may compromise your safety.

Rhonda

For example, Rhonda comes from a home where everyone kisses each other at the breakfast table in the morning and hugs and kisses each other before they go to sleep at night. She is very comfortable with being hugged and touched even by distant relatives whom she sees infrequently. Rhonda works in a jewelry and coin shop after school. Her responsibilities are to "grade" the coins and to polish the sterling silver pieces. She works at a small table facing the wall. Her coworker is a guy named Steve. One day Steve came up behind her and began to rub her neck. Rhonda was a little surprised, but she did not flinch or tell him to stop because it felt good and her neck was stiff from leaning over the work table.

The following week Steve gave her a hug. This did not feel as good as the neck rub, but she did not tell him to stop because he was a nice guy and she did not want to hurt his feelings. The day after the hug, Steve came up behind her and kissed her neck and tried to touch her breasts. Rhonda pushed her chair back and told him to stop. He did, but he looked a little confused. He asked her what was the problem? The neck rub and hugs were okay; didn't she like him?

In this situation Steve really was a nice guy and did not want to hurt Rhonda. They talked about the communication problem and worked things out. However, not all guys are as nice as Steve. Rhonda, despite her boundary system, does not have to let everyone touch her. She can set her boundaries by verbally communicating them sooner than she did with Steve.

The way to set boundaries with people we know is by using a formula for communicating them. Just as Monica was prepared for how she would behave during a robbery, Rhonda can have a "script" that works in almost any situation. She needs to listen to her internal voice and set a boundary. She could say, "I feel uncomfortable when you hug me; please stop."

Let's take a closer look at that sentence and examine what she is doing. First, Rhonda needs to identify her feelings. This is sometimes difficult, because we don't always know what we are feeling. Rhonda knew that the hug did not feel "right," so she expressed that by saying, "I feel uncomfortable." She did not say, "*You* make me feel uncomfortable." "You" language–rather than "I"–tends to put people on the defensive and gives them something to object to. When you own your own emotions, it is harder for people to object to them. You can't tell someone how she feels.

Next Rhonda identified the behavior that made her feel uncomfortable. "I feel uncomfortable *when you hug me...*" This is important. We cannot assume that people can read our minds. It is difficult to tell people what we feel, but if we don't, it is really hard for them to grasp how they can adapt their behavior so that they don't make us uncomfortable.

Finally, Rhonda gave Steve the solution by saying, "Please stop." It is not enough for Rhonda to identify the problem without offering a solution. The solution is for Steve to stop. Steve was a nice guy and did not want to make Rhonda uncomfortable. He stopped. Most people will respect our boundaries if we give them the opportunity.

In addition to verbally communicating the statement, "I feel uncomfortable when you hug me; please stop," it is important to communicate the same message with your body language and facial expression. What if Rhonda set her boundary with appropriate language but giggled and looked away when she said it? Because her language is saying stop and her body language and facial expression are saying "maybe, stop," Rhonda is being inconsistent or incongruent. It is important that we be congruent with body language and facial expression so that the message is clear.

Body language is very important. Many people believe that as much as 90 percent of communication takes place on a body-language level. Body language is the subtle and sometimes not so subtle way we hold our bodies. For example, how many of you know that a parent is angry at you before a word is said? How many of you know when a close friend is sad or depressed? We can get a lot of information about people just by reading their body language.

With Rhonda and Steve, Steve got the sense that a hug would be okay because Rhonda didn't mind his giving her a neck

Sometimes we have to defend our boundaries.

rub. What if Rhonda had brushed his hand away when he gave her the neck rub? Steve might have understood that gesture to mean that she did not like being touched. However, the fact that Rhonda did not mind the neck rub does not mean that Steve has universal permission to touch her in any manner. We still have the ability to set a boundary at any point during an interaction.◆

chapter 4

Social Conditioning

Another dynamic should be addressed when talking about personal safety skills for teenage girls. This dynamic came up in the scenario with Rhonda and Steve. After Steve gave Rhonda a hug, she knew internally that it didn't feel "right." Learning to pay attention to this internal voice is a good habit. Often young women are socially conditioned not to pay attention to their inner voice. They are raised to put other people's needs above their own. Rhonda did not tell Steve to stop because she was afraid of hurting his feelings; she believed that Steve's feelings were more important than her own.

Identifying our social conditioning is difficult. Social scientists have discovered that young women often lose their ability to communicate their thoughts and feelings. Sometimes not until these women reach their mid-twenties do they regain a sense of their personal voice and begin to express themselves.

Women are taught to be quiet and passive. They are sometimes taught that disagreeing or having strong opinions is unfeminine.

But putting other people's feelings above your own and not being able to express your thoughts and feelings may compromise your ability to stay safe. This doesn't mean that you should be inconsiderate of others. It means that you should pay attention: when do you ignore the internal voice that tells you, "Uh-oh, something isn't right"? What prevents you from saying something about it.

Lizette

Lizette had just moved into a new neighborhood. She was 16 years old,

Some people think they can take what they want. They may cross boundaries in doing so.

bright, and very responsible. She moved with her mom after her parents were divorced. After the divorce, finances were tight, and Lizette did not get an allowance. She decided to start baby-sitting and, being the resourceful girl that she was, put flyers around the neighborhood to let people know she was available.

Her diligence paid off. A couple in the community, Sylvia and Brad, hired her to baby-sit their daughter. Later Brad would drive her home. During these drives Brad was really nice. He asked her about herself and really listened to her answers. He often talked to her like a friend, not a teenager. They got into the habit of stopping in front of her home and talking for a while before she went inside.

After a couple of weeks, Brad began asking her some very personal questions. He asked if she had a boyfriend and what she had done sexually with him. She was a little uncomfortable with the conversation, but she did not want him to think she was a prude who couldn't talk about adult things. He was also so attentive, unlike her father, who was so domineering. So she answered his questions. He then began to kiss her good-bye. At first it was a quick peck on the lips. That seemed okay. Then he began to lengthen the kisses. But she wasn't really sure; she thought she might be overreacting. After all, Brad was married and had a child.

One spring evening Brad changed into lightweight cotton shorts before driving her home. She thought this was a little odd; it wasn't that hot out. When they got to her house, Brad asked if she was wearing a bra. Lizette was shocked; that seemed none of his business. She became flustered and *started to answer him, then decided not to. He told her she was really beautiful and smelled wonderful. She was torn between being flattered and being really uncomfortable. He then took her hand and put it on his lap and asked her if she would touch his penis. Lizette froze. She could barely breathe and did not know what to do. She liked Brad; but not in that way. At that moment the lights went on in her house, and she saw her mother looking out the window. She told Brad her mother was waiting for her and she had better go.*

Lizette felt ashamed and responsible for what Brad had done. She felt unable to deal with the situation and was very confused. She was also concerned that Brad might not stop his behavior. How was she going to continue to work for Brad and Sylvia? She needed the money and liked his daughter and the job. She didn't want to offend him. What could Lizette do?

Lizette's Options

Let's look at all of Lizette's options. First,

Your parents might be able to help you deal with your situation at work.

she can tell her mother. Dealing with adult problems is difficult; you don't have to face them alone. Asking for help can be hard because it may seem to imply that you are incapable of handling it on your own. Although this book is designed to teach you skills to enhance your self-reliance, that doesn't mean that asking for help from a parent or other trusted adult is a wrong solution to the problem.

Talking to a parent or adult might get you some valuable information about what happened in the car. First of all, there is a name for Brad's conduct. It is called sexual assault. Sexual assault occurs when a person (often called the perpetrator) touches another person (often called the victim) in a sexual manner or asks that person to touch him. Sexual assault is never the victim's fault, although she or he may feel as though it is. Lizette did not bring about Brad's inappropriate behavior. His conduct before the sexual assault appeared to be normal and within the limits of acceptable conduct. With a closer look, however, it showed a pattern of increased physical contact and emotional invasion.

This pattern of behavior can be labeled as a boundary violation. Brad's asking personal questions about Lizette's boyfriend and her sexual experience could be characterized as a boundary violation. So could a prolonged kiss on the mouth. Objectively, that may be easy to say, but from Lizette's standpoint it did not seem so obvious. Some of you may also believe that Brad's asking personal questions or kissing Lizette on the mouth is okay. The good thing about using boundaries as a concept for personal safety is that you identify what is an important boundary and you set those limits. Boundaries are flexible. What may not be okay one day may be just fine the next.

In this example, Lizette felt a little confused and uncomfortable with Brad's conduct early on. This should send a warning signal to her to pay attention to what is happening and to address the problem as soon as possible. Lizette was confused about Brad's conduct because she believed him to be a nice guy. She did not want to offend him by telling him that his questions were out of line or his kisses too much physical contact. This is another example of how young women may put somebody else's feelings above their own.

Lizette had another challenge as well. Brad was her employer, and she really needed the job. It is a good strategy to be polite to your boss. However, it is not a good personal safety strategy to ignore those "uh-oh" feelings about his behavior.

Now that Lizette has identified the problem, she can address it. One of the simplest solutions to her problem is to quit.

That would be setting a very clear boundary. If she and Brad had no contact, he could not assault her again. That would mean that Lizette would have to find another job. Getting out of range of potentially dangerous situations is what safety experts call "target denial." It can be stated simply: Don't be there.

What if Lizette does not want to quit? She really needs the money, she likes the job otherwise, and she wants to try to solve this problem. Lizette might ask Brad's wife to drive her home. She might get her mother to pick her up, or ask a friend who has a car. This also would be a strategy of target denial. She should have a back-up driver in case for some reason no one else is available (and it is too far to walk), so that Brad cannot drive her home. This solution may actually work. If it does, great.

Let's assume Lizette's mother doesn't have a car, her friends don't drive, and Sylvia twisted her ankle and can't drive for a couple of weeks. Lizette would be wise to have a talk with Brad, preferably before she gets in the car with him again. Lizette might

Your body language should match your words when you are telling someone to respect your boundaries.

prepare what she is going to say and practice the objections Brad might present. The goal is for her to be ready for the discussion, which should be held in a public place. Finally, Lizette should use clear, directive language with Brad.

This is how the conversation might go. Lizette calls Brad at work and says, "I have something important to discuss with you before you drive me home tonight." Brad agrees. That evening, Lizette stands in the doorway of the house, where she could easily call for help from Sylvia, and says, "Do not touch me again." This is clear, directive language. (What if she said, "I would prefer it if you didn't touch me" or "I don't like it when you touch me"? This is not the time to be overpolite or to make subtle references to what happened. Such a statement might give him the opportunity to say, "Well, maybe if you weren't so uptight sexually you might like it.") This is not really a discussion in the sense of trying to reach a compromise. No compromise is possible. Brad's behavior needs to change, quickly and dramatically.

It is also important that Lizette be congruent with her body language and facial expression. This is a serious situation and needs to be addressed as such. The tone of Lizette's voice should be firm and level, as you would speak to a dog or a young child.

Lizette may or may not be able to address this situation with Brad and get him to change his behavior. There are no guarantees or absolutes in personal safety. If she ignores the situation, however, it probably will get worse. Ignoring problems or pretending that nothing is happening is potentially dangerous. The decision should not be, "Do I address this problem?" but "How can I address this problem?" It is important to face these problems earlier rather than later. They tend to escalate. Stopping boundary violations early can often prevent more serious ones. It lets the perpetrator know that you are not an easy mark.

Brad's behavior—sexually assaulting a teenager—is a crime. If he does not stop when Lizette tells him to, she should tell her mom, Sylvia, or even the police.♦

Chapter 5

Dealing with Sexual Harassment

Sexual harassment comes in many shapes and forms. Although it is not a new phenomenon, it has gathered a lot of recent media attention. There are two basic types of sexual harassment: *quid pro quo* and hostile environment. *Quid pro quo* is a Latin phrase used to describe a situation in which an employee is asked to perform sexual favors in order to receive a promotion or keep her job. Hostile environment refers to a situation in which the workplace is so unfriendly that it makes it virtually impossible for the employee to do her job.

No matter what form sexual harassment comes in, it is illegal.

Barbara

For example, Barbara works in the electronics department of a department store on weekends. Part of her pay is a commission based on her sales. During the last few months she has outsold every salesperson, even though she works only part time. She decides to talk to her boss, Perry, and ask for a raise. Perry tells her that her work is outstanding and he would consider giving her a raise if she would come sit on his lap and give him a big hug. This is an example of quid pro quo *sexual harassment*. It is illegal.

Every state has slightly different laws on filing sexual harassment claims. The basic steps to enforcing your rights generally include reporting the incident at your place of employment, filing a claim with the Equal Employment Opportunity Commission (EEOC), and/or starting a law suit in state or federal court.

If Barbara's coworkers also had posted pictures of naked women in the back room where they stored equipment, constantly made remarks about her body, her dress, her hair, and makeup, and belittled her in front of customers, Barbara might also have a claim for hostile environment. Many corporations are trying to educate their employees and create policies to prevent this type of sexual harassment before it begins. Barbara's legal recourse is much the same as with a *quid pro quo* claim.

It is important that sexual harassment be taken seriously. Even though the examples given did not immediately threaten Barbara's safety, they are situations that need to be addressed. Her boss may try to retaliate for her refusal to sit on his lap, and the hostile environment is already a challenge for her to work in. Often this kind of behavior is tolerated because women are afraid of losing their job or inciting even more harassment. There is no magic way of dealing with this sexual harassment. However, not dealing with it does not make it go away. In fact, it may even make the situation worse.◆

chapter 6

Workplace Violence

Bettina works the evening shift in a shopping mall. All the employees are required to park in the farthest lot from the mall exit/entrance. Most evenings Bettina walks to her car with a couple of coworkers. This Saturday night the store manager is off, however, and she was asked to cash out the register and close up the store. She was thrilled to be given more responsibility and wanted to show that she could handle the job. She was hoping to get a promotion to assistant manager for the summer when she could work full time. Bettina knew that she would have to go to her car alone, but she did not want to pass up this opportunity to prove herself, so she agreed to work late.

As Bettina left the store she noticed how truly deserted the parking lot was. She started toward her car. Feeling a presence behind her, she turned around but could see no one. She told herself she must be paranoid. Bettina continued to walk, but she could not shake the feeling that she was being watched. This time she heard footsteps, and when she turned around she saw a man. Bettina did not know what to do. Should she run? If so, where to? Should she scream? The place seemed so deserted. She could feel her heart pounding. The man approached her and asked if he could walk her to her car.

She said no thank you, that she was fine on her own, but he became more insistent and reached for her arm. Bettina quickly swung her bag at his face. The strap caught his eye. She kicked him in the shins with her cowboy boots. He turned and hobbled away. She ran to her car and drove off.

Bettina's Options

Bettina was able to protect herself in a

You may put yourself in danger by working late or in a deserted area.

dangerous situation for a number of reasons. First, she acknowledged that walking to her car in a deserted parking lot was dangerous. She had choices to make about whether or not to do this. Working late may gain the favor of your boss, but it also might compromise your safety. Bettina might have agreed to work late if one of her coworkers stayed with her, or she might have asked whether mall security would escort her to her car. Perhaps she could have made arrangements to be picked up or gotten special permission to park her car in a better spot.

None of this is to imply that Bettina should not have accepted the manager's offer to close the store. However, to the extent that our routine changes, it is wise to consider how that change might affect our personal safety. If Bettina could not make any other arrangements for that evening, the next best thing would be to create an emergency plan. Being prepared for situations increases our chances of success.

In this situation, Bettina acknowledged that her routine had changed and she needed to be prepared. Before she left the store she called her parents and let them know that she should be home in about half an hour. Even though she was tired, she made sure to shake it off and heighten her level of awareness. She looked both ways before leaving the mall and took a long sweeping look around the entire parking lot. As she walked to her car, she continued to look 360 degrees around her. This kind of surveillance serves two purposes. First, if there is a problem you can see it happening farther away; being surprised gives you less opportunity to address the situation. Second, this behavior makes you look less "like a victim." Just as body language communicates a lot during conversations, it also communicates a lot to would-be criminals. Criminals are masters of sizing up people as to how compliant they may be. Walking with alertness and purpose makes you look less like a possible victim. This is sometimes called a "nonvictim attitude."

In addition, Bettina needed a plan. Awareness is a great tool, but it does not guarantee safety. Using verbal skills to deescalate or set boundaries is a good option for maintaining personal safety. For example, when Bettina saw the guy following her, she could have said, "Leave me alone. Stop following me. I don't want any problems." This is an example of using directive language to set a clear boundary. Being verbally loud has been demonstrated to be one of several behaviors associated with avoiding rape. An article entitled "The Effects of Resistance Strategies on Rape" (*American Journal of Public Health*, November, 1993) reported that forceful verbal resistance, physical resistance, and flight were all associated with rape avoidance, whereas nonforceful verbal resistance and no resistance were associated with being raped.

Bettina might also have lied to the guy. "Stop following me. The security guards in the mall are watching me. They can see that you have approached me and will be here any second." The "right" thing to say may vary because each situation is different and every person has different skills and abilities.

If verbal skills don't work and you think that you are being abducted or assaulted, using physical skills may make sense. Many people hold that fighting back may anger your attacker and make the situation worse. New data suggest, however, that fighting back does not increase your chances of being hurt or the nature or severity of physical injury. The researchers found that women who used forceful resistance were no more likely to be injured than women who did not resist.

If you are going to make a physical re-

Taking self-defense classes is one option for learning how to stay safe.

sponse, however, the following has a good chance of success. None of you will become experts after reading it, so you would be wise to take a good self-defense program to gain skills in this area. However, a little knowledge can go a long way.

Strike target areas only. Target areas are vulnerable on attackers, even strong, large men. Eyes, throat, nose, temples, groin, knees, shins, and insteps. If you are holding an object, throw it into one of those target areas. Commit yourself to fighting 100 percent and fighting for your life. If one thing does not work, try another. Yell loudly. Describe what is happening if you think you might be able to attract attention. Yell something like, "There's a man in a blue shirt and denim jeans, wearing a leather jacket. He is assaulting me. I am in the parking lot in the county mall. Call 911."

Often people who might come to your aid stay away because they don't know what is happening, whether you want help, or how to help. Letting them know that the person is a stranger and is trying to abduct you, describing him, and saying what you need is a way to help them help you. Asking for help by alerting them to the fact that you are in trouble gives them permission to get involved.

In the self-defense program that I teach, Prepare Self-Defense, we teach children, teens, and adults how to fight back in realistic, dynamic scenarios that recreate common attack situations. Prepare believes that you learn best by doing. We teach verbal and physical skills with mock attackers who wear 40 pounds of protective armor. In that way, we can do role plays with people who look threatening and we can learn physical skills by actually hitting a person who is fully protected.

What works so well about this type of training is that you learn to fight in an adrenalized state. This is that heart-pounding feeling you get when your body is preparing to fight or flee. We also train how to break through the freeze response, which is a natural biological response to fear or danger, and then think and react appropriately. A deer may freeze in its tracks when it believes it is being hunted. A deer does that because its coat helps it blend into the environment. However, humans can't camouflage themselves when spotted by a predator. Freezing when confronted with a dangerous or scary situation doesn't keep us safe.

By training in this type of dynamic scenario, we can train students to shorten their freeze response, start making assessments of exactly what is happening, and choose the best response in that situation. No system of training is perfect, and there are no guarantees; however, the more training and skills you have, the more likely you are to get out of the situation unharmed. Studying martial arts is another option. Make sure to shop around and speak to other teens who are studying at a particular school. Try to get a sense that the school addresses the types of assault that women are subjected to, and understand that techniques that work best for men, who have upper-body strength, may not work as well for women, who have more strength in their legs. Understand that it takes many years to become proficient in a martial arts style, and, even after many years of practice, what you learn in the school may not apply to street scenarios.◆

Conclusion

As we grow up, work and our work environment become a larger part of our everyday life. Unfortunately the work setting is not as safe as we would like. Robberies, assaults, and sexual harassment are problems that we can address better if we take the time to get prepared.

Being aware and learning how to identify dangerous situations and avoid them when we can is the best way to stay safe. You can get prepared by taking inventory of your work environment and figuring out where in that environment danger exists. Pay attention to any "uh-oh" feelings, and don't be afraid to speak up and set boundaries. If

Your work environment should be a safe, comfortable place.

you are in a life-threatening situation and decide to make a physical response, strike with 100 percent of your power to target areas. Don't stop until you know you can get to safety. Fighting spirit may be even more important than fighting technique. If you want more skills, take a self-defense class.◆

Glossary

adrenalized state See fight/flight syndrome.

assailant Person who commits a crime.

body language The way we hold our body and the messages and information that communicates to others.

boundary The physical distance or emotional limit that surrounds a person.

boundary violation An event in which a person invades your comfort zone with either physical touching or verbal intimidation or disregard for your emotions.

comfort zone Your own personal physical boundary.

deescalation Using verbal skills to intervene and stop or prevent a crime from beginning or continuing.

directive language Words that clearly state what you want, *i.e.*, Stop, Go away, Leave me alone, Back off.

fight/flight syndrome Biological response in which adrenaline and other hormones are released into the bloodstream, enabling the person to fight or run away.

freeze response Biological response to fear or danger that can paralyze a person.

hostile environment Type of sexual harassment that refers to work site behavior or environment that makes it impossible for an employee to do her job.

internal voice Your own thoughts and feelings.

nonvictim attitude Body language, facial expression, and verbal traits that make you look confident and assured.

perpetrator Person who commits a crime.

quid pro quo Type of sexual harassment in which an employee is asked for sexual favors in return for keeping or advancing on a job.

sexual harassment Behavior that includes unwanted touching or verbal intimidation at the workplace.

victim The target of a crime.

Resource List

General Information

Women's Bureau of the U.S. Department of Labor (through regional offices women can obtain information about job training, receive advice about discrimination, find referrals, and obtain educational and statistical data about women; see regional offices)

ACLU Women's Rights Project
American Civil Liberties Union
132 West 43rd STreet
New York, NY 10036

Equal Employment Opportunity Commission
800-669-4000

Equal Rights Advocates
1663 Mission Street
San Francisco, CA 94103
415-621-0672

National Organization for Women
1000 16th Street NW
Washington, DC 20036
202-331-0066

9 to 5, National Association of Working Women
614 Superior Avenue NW
Cleveland, OH 44113
800-245-9865

Women Plan Toronto
736 Bathurst Street
Toronto, Ontario M5S 2R4
416-588-9751

Information on Crime and Victimization

Crime Victims Counseling
P.O. Box 023003
Brooklyn, NY 11202-0060

National Organization for Victim Assistance
1757 Park Road NW
Washington, DC 20010
800-879-6682

In Canada, call 1-800-VICTIMS (1-800-842-8467)

Rape Crisis Centers (for a nationwide listing of rape crisis centers, call the Washington, DC, Rape Crisis Center Hotline, 202-333-7273, or check the phone book for local information)

Ottawa Sexual Assault Support Centre Hotline
613-234-2266

Toronto Rape Crisis Centre Hotline
416-597-8808

Hospital Emergency Room (Ask for Rape Trauma Center)

The Police (For emergencies dial 911; check the phone book for local information)

Information on Self-Defense Training

Impact Personal Safety
19310 Ventura Boulevard
Tarzana, CA 91356
818-757-3963

Prepare Self-Defense
25 West 43rd Street
New York, NY 10036
800-442-7273

Woman's Way Self Defense
512 Silver Spring Avenue
Silver Spring, MD 20910

The YWCA and Martial Arts
(Check the phone book)

In Canada, call Impact Personal Safety at 818-757-3963 for references to Canadian self-defense programs

For Further Reading

Bravo, Ellen, and Cassedy, Ellen. *The 9 to 5 Guide to Combatting Sexual Harassment.* New York: John Wiley & Sons, 1992.

Brownmiller, Susan. *Against Our Will: Men, Women, and Rape*, rev.ed. New York: Bantam Books, 1988.

Caignon and Groves. *Her Wits About Her.* New York: Harper and Row, 1987.

Cooney, Judith. *Coping with Sexual Abuse*, rev.ed. New York: Rosen Publishing Group, 1991.

Gilligan, Carol. *In a Different Voice: Psychological Theory and Women's Development.* Cambridge: Harvard University Press, 1982.

Langelan, Martha. *Back Off! How to Confront and Stop Sexual Harassment and Harasser.* New York: Simon & Schuster, 1993.

Index

A
ATM, 6
attention, attracting, 51
awareness, 8, 10–13, 48, 54

B
baby-sitting, 30–32
body language, 17, 26, 39, 48
boundaries
  knowing, 19–27
  setting, 24, 26–27, 36, 48, 54
  unclear, 20
  violations of, 34, 39

C
claims, sexual harassment, 42
comfort zone, 20, 22
communication, verbal, 25–26
conditioning, social, 28–39
congruence, 39
coworkers
  bigoted, 12–13
  mental checklist of, 11
crime, property
  escalation of, 17–18

D
directive language, 38, 48

E
emotions, owning one's own, 25
environment
  attention to, 10
  outside workplace, 11

F
feelings, identifying, 24
freeze response, 16, 32, 52

H
hostile environment harassment, 40, 42, 43
hugging, unwanted, 22, 23, 25, 26–27

I
incongruence, 26
"I" vs. "You" language, 25

K
kisses, unwanted, 23, 31, 34–35

L
limits, setting, 35
lying, as defense measure, 49

M
martial arts, 53

N
nonvictim attitude, 48

O
orders, obeying robber's, 16–17

P
parent, help from, 33
passivity, 29
physical skills, 49–50, 51, 56
plan
  safety, 14–18, 47, 48
Prepare Self-Defense, 51–52
problem
  identifying, 35
  ignoring, 39

Q
questions, personal, 31, 35

quid pro quo harassment, 40, 42, 43

R
rape, 17, 19
robbery, 6, 14, 16, 24, 54
role-playing, 52

S
safety
  compromised, 29, 46
  responsibility for, 8
  skills, 28, 35
security arrangements, 11, 14–15, 46
self-reliance, 33
sexual assault, 19, 33–34, 39, 54
sexual harassment, 40–43, 54
situation, dangerous
  assessing, 8, 52, 54

deescalating, 8

T
target areas, 50–51
target denial, 36
touching, unwanted, 6, 22, 34

U
"uh-oh" feeling, 29, 35, 54

V
verbal skills, 48–49, 51
voice, internal, 24, 28

W
workplace
  assessment of, 54
  mental checklist of, 11–12, 15–16
  violence in, 44–53, 54

Acknowledgments

This book is dedicated to Karen Chasen, the Executive Director at Prepare Self-Defense. Without her constant support, editorial eye, and sense of humor, none of these books would have been completed. Many other people have taught me about self-defense and personal safety. Listing them all would take many pages. However, my deepest thanks go to Lisa Gaeta, Director of Impact Personal Safety, Los Angeles. Lastly, I want to thank my parents who have always taught me to "be aware."

About the Author

Donna Chaiet, a practicing attorney in New York City, is the founder and President of Prepare, Inc. Prepare conducts personal safety programs that teach teenagers the verbal and physical skills required to defend themselves by training them to fight against a padded mock assailant. Ms. Chaiet is a recognized speaker and conducts safety/communication seminars for schools, community organizations, and Fortune 500 companies throughout the United States. Ms. Chaiet's frequent television appearances include CBS, NBC, ABC, WOR, FOX, Lifetime, Fox Cable, and New York 1.

Photo Credits

cover by Michael Brandt; photos on pp. 12, 18, 37 by Katherine Hsu; all other photos by Kim Sonsky

Design

Kim Sonsky